Blood Gases

Made Simple, Easy and Quick

Donald A. Thompson, RN, MSN, FNP

Lulu Press, Inc.

ISBN 978-0-6151-6555-4

Lulu Press, Inc.

Learning Objectives

When you have finished this book you will be able to:

1. Briefly define the following parameters:
   - pH
   - $pCO2$ (p_aCO2)
   - $HCO3^-$ (bicarbonate)
   - $pO2$ (p_aO2)
   - S_aO2 (oxygen saturation)
   - S_pO2
   - EtCO2 (end-tidal carbon dioxide)
   - BE (base excess).

2. List normal values and ranges for the above parameters.

3. Define the following conditions and indicate their associated lab values: acidemia, alkalemia, metabolic acidosis, respiratory acidosis, metabolic alkalosis, respiratory alkalosis, hypoxemia.

4. Describe the process of compensation.

5. State the major clinical significance of abnormal anion and osmolar gaps.

6. List normal values for anion and osmolar gaps.

7. Recognize a normal capnogram (EtCO2) tracing.

8. Interpret any given set of arterial blood gas (ABG) values.

An arterial blood gas (ABG) set consists of 3 parameters:

pH

pCO2 (partial pressure of carbon dioxide)

HCO3⁻ (bicarbonate)

In interpreting an ABG set, it is best to inspect each of these parameters in the above order.

Associated parameters are:

p_aO2 (partial pressure of oxygen)

S_aO2 (oxygen saturation)

BE (base excess)

EtCO2 (end-tidal carbon dioxide)

pH

> pH = a measure of the acidity or alkalinity of a solution (hydrogen ion [H^+] concentration)

In an ABG set, the first parameter you want to look at is the pH.

The normal value for the pH of human blood is 7.4.

The body's ability to maintain this precise value of 7.4 is remarkable, that is, the body keeps the pH very "tightly wrapped" around 7.4.

We often express the normal pH as **7.40**. ⇦ *Note the trailing zero.*

Mathematically 7.4 and 7.40 are the same number, but when we speak of the pH, we use the zero at the end to emphasize the precision with which the human body maintains the pH.

The normal range for the pH is commonly given as 7.35 – 7.45.

This is the range used in this book, but you will find that some authorities use an even narrower range around 7.40 as the normal range for the pH.

For example, some references give 7.38 – 7.42 as the normal range.

Whatever range you use as the normal range, bear in mind that even a slight deviation of the pH in either direction from 7.40 should make you suspect a possible acid-base disorder.

On the next page, carefully look over the graph of the pH scale.

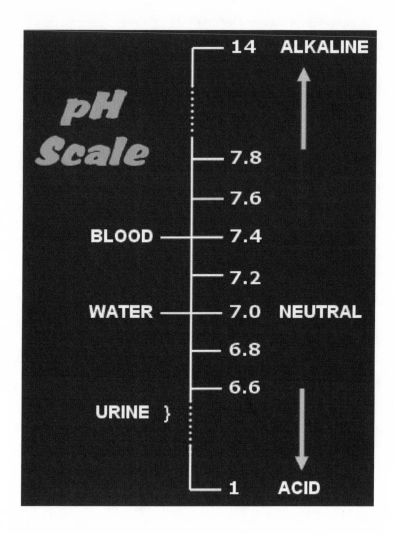

Yes, the pH of normal human blood is slightly alkaline (7.4).

If the pH deviates in a direction *below* the normal range (less than 7.35), we say that the blood is (too) *acid* and we call this condition *acidemia*.

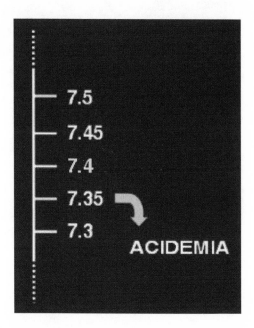

If the pH deviates in a direction *above* the normal range (greater than 7.45), we say that the blood is (too) *alkaline* or *basic*. This condition is termed *alkalemia*.

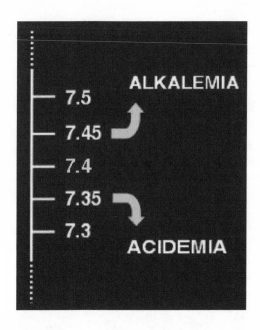

The pH is referred to as the *acidemia-alkalemia* parameter of the ABG set.

Summarizing…

ABG Set	
pH	**acidemia-alkalemia** parameter
pCO2	
HCO3⁻	

pCO2

The second parameter you want to inspect is the pCO2.

The pCO2 is the *respiratory* parameter.

Stated most precisely, the pCO2 reflects the status of the client's *ventilation*.

Summarizing…

ABG Set	
pH	acidemia-alkalemia parameter
pCO2	**respiratory** parameter
HCO3⁻	

The abbreviation "pCO2" stands for "partial pressure of carbon dioxide."

Sometimes you will see it abbreviated this way:

P$_a$CO2

The subscripted "a" stands for "arterial." However, for the sake of brevity the "a" is commonly omitted in everyday parlance.

The normal (mean) value for pCO2 is 40 mm Hg.

If the alveoli are being cleared of carbon dioxide properly, the pCO2 will be normal (namely, the pCO2 will be 40).

If the alveoli are *retaining CO2* (because of poor respiratory movements), CO2 will dam back in the blood and the pCO2 will rise above 40.

This accumulation of CO2 in the blood is an *acidosis* because, when CO2 dissolves in water, it forms a weak acid known as carbonic acid. You may remember this equation from your basic chemistry class:

$$H_2O + CO_2 \Rightarrow H_2CO_3$$

(carbonic
acid)

The higher the pCO2 rises above 40, the more acidic the blood becomes, and consequently, the more the pH *drops*.

Recall that on the pH scale, the smaller numbers are at the acidic end, and the larger numbers are at the basic end.

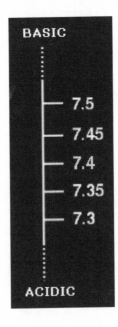

Whenever there is an excessive level (partial pressure) of CO_2 in the blood, a *respiratory acidosis* is said to exist.

(Note: The terms `acidemia` and `acidosis` are not quite the same thing. This will be explained shortly.)

If there are excessive respiratory movements (hyperventilation), the lungs "blow off" CO_2, which decreases the pCO_2. Consequently, there is less carbonic acid in the blood and the pH rises (shifts toward alkaline).

This decreased level (partial pressure) of CO_2 in the blood resulting from hyperventilation is termed *respiratory alkalosis*.

Note that a high pCO_2 defines a *respiratory acidosis* and that a low pCO_2 defines a *respiratory alkalosis*.

Defining…

⬆ pCO2 = Respiratory Acidosis

⬇ pCO2 = Respiratory Alkalosis

The normal range for the pCO_2 is 40 plus or minus 5, that is, 40 ± 5 .

In other words, the pCO_2 may normally vary between 35 and 45, but the mean (average) value is 40.

12

To help you learn the normal blood gas values, note the similarity in the values for both pH and pCO2:

	Mean Value	Normal Range
pH	7.40	7.35 – 7.45
pCO2	40	35 – 45

HCO3⁻

The third parameter of the ABG set is the bicarbonate level or HCO3⁻.

The HCO3⁻ is the *metabolic* parameter.

The HCO3⁻ reflects whether there is a derangement (acidosis or alkalosis) of metabolic origin.

Summarizing…

ABG Set	
pH	acidemia-alkalemia parameter
pCO2	respiratory parameter
HCO3⁻	**metabolic** parameter

The normal value for the HCO3⁻ is 24 mmol/L.

The bicarbonate concentration normally varies from the mean value of 24 by only 2 in either direction, that is, 24 plus or minus 2 (24 ± 2).

Stated differently, the reference range for HCO3⁻ is 22 – 26.

Bicarbonate is a base (alkali).

If there is more base in the blood than normal, that is, if the HCO3⁻ is elevated, we say there is a *metabolic alkalosis*.

This will tend to make the pH *rise*.

14

If some process causes the HCO3⁻ to decrease below the normal range, we say there is a *metabolic acidosis*.

This will tend to make the pH *drop*.

Anion = negatively charged ion in an electrolyte solution.

An example of metabolic *alkalosis* can occur with the old-fashioned remedy of taking baking soda for "acid indigestion." Baking soda is sodium bicarbonate, and if too much of the bicarbonate anion is ingested, metabolic alkalosis results. This is why the labeling on a package of baking soda warns against using it as an antacid.

An example of metabolic *acidosis* is DKA (diabetic ketoacidosis). In DKA, the acid ketone bodies which accumulate in the diabetic's blood consume the bicarbonate anion.

Note that an elevated HCO3⁻ defines a *metabolic alkalosis*, and a depressed HCO3⁻ defines a *metabolic acidosis*.

Defining...

⬆ HCO3⁻ = Metabolic Alkalosis

⬇ HCO3⁻ = Metabolic Acidosis

Acid-Base Imbalances

When speaking of ABG's, the ending **-emia** (as in "acidemia") refers to the pH.

The ending **-osis** (as in "acidosis") refers to the disturbance (process) that is taking place in either the pCO2 or the HCO3⁻.

Thus, if the pCO2 is 55 (high), we know that there is an acidotic process taking place (specifically, a respiratory acidosis).

If the HCO3⁻ is 19 (low), this also is an acidosis (in this case, metabolic in origin).

You may be wondering at this point whether it is possible to have both respiratory acidosis and metabolic acidosis in the same individual at the same time.

Yes, both types of acidosis can indeed co-exist in the same person, a condition called *mixed acidoses*.

Also, as we will see later, one can have *mixed alkaloses*.

As we will see shortly in the discussion on compensation, the pH does not necessarily have to be acidemic even though there is an acidosis present.

Likewise, if an alkalosis is present, the pH does not necessarily have to be alkalemic.

The terms "acidemia" and "alkalemia" refer to the *measured acid-base level* of the blood.

"Acidosis" and "alkalosis" refer to *disorders* (processes) that have the effect of *tending* the blood toward acidic or basic.

Terminology of ABG Set		
Acid*emia* -or- Alkal*emia*	Acid*osis* -or- Alkal*osis*	
pH	pCO2	HCO3⁻

Summarizing...

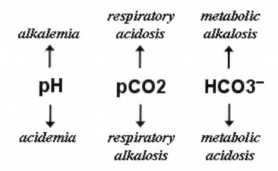

The arrows in the above diagram represent movement away from the normal value. Thus, if the pCO2 is high, it is termed "respiratory acidosis."

Oxygen

Ventilation (air moving in and out of the airways) and oxygenation (the amount of oxygen being transported in the bloodstream) are best considered as separate entities. A change in one may affect the other, but there is not necessarily a one-to-one correlation.

Recall that the *ventilation* parameter is the pCO2.

The two major parameters for assessing the status of a patient's *oxygenation* are:

- Partial pressure of oxygen in arterial blood (p_aO2 or pO2 or oxygen tension)

- Oxygen saturation in arterial blood (S_aO2 or SO2 or "O2 sat").

Of all the oxygen in the blood, 95% travels as *oxyhemoglobin* (oxygen-saturated hemoglobin), and 5% is dissolved in the plasma. The portion dissolved in the plasma is the pO2.

S_aO2 is the percentage of oxyhemoglobin in the total arterial hemoglobin pool. This is shown in the formula below:

$$S_aO2 = \frac{oxyhemoglobin}{oxyhemoglobin + reduced\ hemoglobin} \times 100$$

Reduced (desaturated) hemoglobin is the hemoglobin that does not have oxygen molecules attached.

If some condition prevents the hemoglobin from being "loaded" with oxygen, the S_aO2 will be low.

The S_aO2 and p_aO2 are related, but the relationship is non-linear. A sampling of arterial blood is needed for a p_aO2 reading, but a pulse oximeter can obtain the S_aO2 non-invasively.

The oximeter probe that is attached to the client's fingertip or earlobe has two sides: an *emitter* and a *sensor*.

The emitter sends red and near-infrared light into the skin. The light hits red blood cells and is partially absorbed by hemoglobin.

The amount of light absorption depends on the degree to which the hemoglobin is saturated or desaturated with oxygen.

The part of the light that is not absorbed passes through the tissue and is picked up by the sensor. The microprocessor chip in the monitor analyzes the non-absorbed light to determine the amounts of saturated and desaturated hemoglobin.

A pulse oximeter is used to measure oxygen saturation in the peripheral capillaries, so this measurement is most accurately called the S_pO2, where "p" stands for "peripheral."

In health, the saturation of *peripheral* oxygen (S_pO2) should be nearly identical to the saturation of *arterial* oxygen (S_aO2), so oximetry saturations are often called S_aO2. However, S_pO2 is the more precise term for a saturation measurement done with an oximeter.

Every health professional should know the following normals:

Abbreviation	Name	Normal
p_aO2	Partial pressure of oxygen	75 – 100 mm Hg
S_aO2	Oxygen saturation	94 – 100 %

An S_aO2 value less than 90% is *hypoxemia*.

S_aO2 and pO2 often approximate each other in the clinical setting, but they are not interchangeable. They measure different things (S_aO2 measures oxygen attached to hemoglobin; pO2 measures oxygen dissolved in plasma). Note the different units, mm Hg vs. %.

You may sometimes see the results of a blood gas measurement recorded in progress notes like this:

7.4 / 40 / 98 / 24 / 99

pH / pCO2 / pO2 / HCO3- / S_aO2

Base Excess

Sometimes, especially in pediatrics, you will see an additional forward slash and number tacked on to the right end of the train of values shown on the previous page. This is the base excess (BE).

BE gives us information about the *metabolic* acid-base status of the patient.

BE is the amount of base that must be added to restore a normal pH. It is determined from an equation that uses the values for $HCO3^-$ and pH.

The normal range of BE is -2 to +2 mEq/L.

If BE is high (greater than +2), there is said to be an "excess of base" which confirms a metabolic alkalosis.

If BE is low (less than -2), there is said to be a "deficit of base" which confirms a metabolic acidosis. If BE is below -10, a severe metabolic acidosis is present.

EtCO2

End-tidal carbon dioxide (EtCO2) is the level of carbon dioxide at the very end of expiration, and is an indicator of the status of *ventilation*. It correlates closely to the pCO2.

However, the pCO2 requires a specimen of arterial blood. An EtCO2 monitoring device can determine the EtCO2 non-invasively by analyzing a sample of expired gases.

EtCO2 measurement has many applications in the clinical setting, but an area where one can readily appreciate its usefulness is in checking endotracheal tube placement. If an endotracheal tube is inserted into the esophagus, the EtCO2 monitor, as you can imagine, will detect little or no CO2.

Capnography, or the study of EtCO2 *waveforms*, yields valuable information about the patient's respiratory condition (emphysema has a characteristic pattern, for example) or problems with the ventilation equipment (such as leaks in the system).

A *capnogram* is an EtCO2 waveform tracing. A *capnograph* is a machine that contains a CO2 sensor and a screen and/or printer to display the waveform.

Here is a sample normal capnogram:

Note the "square box" shape of each waveform with the right limb slightly higher than the left.

The highest point on an EtCO2 waveform (the "peak" on the right limb) indicates the patient's EtCO2 value (also called $P_{et}CO2$). In a normal person, the $P_{et}CO2$ is very close to (slightly less than) the CO2 level in arterial blood (pCO2).

Thus, the normal range given for the $P_{et}CO2$ is the same as that for the pCO2, namely, 35 – 45 mm Hg.

Compensation

When an acid-base disturbance (i.e., acidosis or alkalosis) has been present for some time, the body will attempt to *compensate* by generating the opposite process.

Thus, if there has been an acidosis for a long time, the body may "kick in" an alkalosis. It may even succeed in bringing the pH back to the normal range (although it is virtually impossible for the pH to swing all the way back to a perfectly normal 7.40).

This phenomenon is called *compensation* or *correction*.

Let's say there is a metabolic acidosis which initially causes the pH to drop to 7.2. If this continues for any length of time, the body will try to compensate by increasing the respiratory rate, even to the point of frank hyperventilation.

The rapid respirations cause the lungs to blow off CO_2, which decreases the pCO_2.

Since a decreased pCO_2 defines a respiratory alkalosis, we say that a *compensatory respiratory alkalosis* is occurring.

The compensatory alkalosis will counteract the acidity of the blood, so the pH will rise from its low position of 7.2.

The pH may even rise until it is within the normal range, but it will probably not be completely corrected to 7.40.

To compensate for a primary acid-base disorder, the body sets up an opposite process. Thus, a primary alkalosis is compensated for by a secondary acidosis and vice-versa.

If compensation is just beginning and the pH has not yet moved back into the normal range, we say that the primary disturbance is only *partially* compensated.

If the pH has reached the normal range, we say that the primary disorder is *fully* compensated.

Let's consider the following set of ABG's:

pH	pCO2	HCO3⁻
7.20	40	15

We look at the pH first and see that it is low and that the blood is therefore acidemic.

The acidemia is undoubtedly the result of an acidosis, so we next look at the pCO2 and HCO3⁻, in that order, to see where the acidosis is coming from, that is, whether it is respiratory or metabolic.

We see that the pCO2 is normal and the HCO3⁻ is low. Thus, the source of the acidemia can only be the HCO3⁻.

Since the HCO3⁻ is the metabolic parameter and it is low (less basic), this has to be a metabolic acidosis.

Note that the pCO2 is completely normal, so there is no compensation.

(If there were compensation here, the pCO2 would have to decrease.)

Thus we have the interpretation: uncompensated metabolic acidosis.

It can sometimes happen that you have an acidemia which is brought about by both a respiratory acidosis and a metabolic acidosis. This is called *mixed acidoses*.

The following is an example:

pH	pCO2	HCO3⁻
7.18	50	18

We look at the pH first and see that it is markedly acidemic.

Second, we note that the pCO2 is high and that by definition a respiratory acidosis is present.

Finally, we observe that the HCO3⁻ is low and that therefore a metabolic acidosis is also occurring.

We can also have *mixed alkaloses* as in this example:

pH	pCO2	HCO3⁻
7.62	30	30

First, note that the pH is intensely alkalemic.

Second, we see that the pCO2 is low. By definition this is a respiratory alkalosis.

Third, the HCO3⁻ is high, so we have a metabolic alkalosis, too.

Thus, both respiratory and metabolic sources contribute to the alkalemia.

Now let's return to our original case of *metabolic acidosis*:

pH	pCO2	HCO3⁻
7.20	40	15

To recap, the pH is acidemic, the pCO2 is normal, and the bicarbonate is low (the latter indicating a metabolic acidosis). Since the pCO2 is normal, no respiratory compensation has (yet) taken place, so we say that this is an uncompensated metabolic acidosis.

Let's say that this is a diabetic in acute ketoacidosis (DKA). After a time, the body will try to correct the acidosis by blowing off CO2. For example, we see this in the "air hunger" (Kussmaul's respirations) which occurs in prolonged DKA.

The hyperventilation will lower the pCO2, thereby setting up a compensatory respiratory alkalosis.

So, we started out with an *uncompensated* metabolic acidosis.

Later, a secondary respiratory alkalotic process began to compensate:

pH	pCO2	HCO3⁻
7.30	32	15

The pCO2 has fallen below the normal range, indicating that there is a respiratory alkalosis going on. The pH has risen somewhat, but it is still acidemic. We say that this is now a *partially compensated* metabolic acidosis.

As compensation continues, the pH rises into the normal range:

pH	pCO2	HCO3⁻
7.37	27	15

To review, our metabolic acidosis was initially *uncompensated*:

pH	pCO2	HCO3⁻
7.20	40	15

Then the metabolic acidosis became *partially compensated*:

pH	pCO2	HCO3⁻
7.30	32	15

In the ABG set above, note that the:
- bicarbonate has not changed.
- pCO2 is below normal.
- pH is less acidemic, but still not back to the normal range.

In the set below, the respiratory alkalosis has *fully compensated* the metabolic acidosis. Note that a perfect 7.40 has not been attained.

pH	pCO2	HCO3⁻
7.37	27	15

Ambiguous Values

Occasionally you may run into a set of ambiguous values.

For example, consider this ABG set:

pH	pCO2	HCO3⁻
7.42	52	33

This set of values is compatible with 2 possible interpretations:

- fully compensated metabolic alkalosis; and

- fully compensated respiratory acidosis.

The best clue as to which of the above 2 disorders would be most appropriate for a given patient, of course, is the total clinical picture.

However, note that the pH in the above case, even though in the normal range, is "leaning" toward the *alkali* side.   This makes it more probable that the actual interpretation is fully compensated metabolic *alkalosis*.

This approach to interpretation of ambiguous values is followed throughout this text.   That is, whenever an ambiguous set of values is encountered, the interpretation that is given is the one suggested by the direction *toward which the pH is tending*.

4-Step Method
for ABG Analysis

The key to ABG interpretation is the pH!

1. Look at the pH. Note whether it is:

    a. Perfectly normal (= 7.40).

    b. Distinctly abnormal (below 7.35 or above 7.45).

    c. Low normal (7.35 – 7.39) or high normal (7.41 – 7.45).

2. If the pH is **perfectly normal**, odds are that the other 2 parameters are also normal (if they are, you have identified a set of normal ABG's).

3. If the pH is **distinctly abnormal**:

    a. Note whether it is acidemic or alkalemic.

    b. Look at pCO2 and $HCO3^-$ to identify a process corresponding to the pH abnormality (alkalosis if pH is alkalemic, acidosis if pH is acidemic). If so, this is the primary disorder:
    — if in the pCO2, it is *respiratory*.
    — if in the $HCO3^-$, it is *metabolic*.

    c. Look at the remaining parameter to see if there is any **compensation**. If this last parameter is normal, the primary disturbance is *uncompensated*. If there is compensation, the primary disorder is *partially compensated*.

4. If the pH is **low normal or high normal**, the other 2 parameters may also be normal, but this is a clue that there may be a *fully compensated* disturbance.

   a. Note the direction toward which the pH value tends, i.e., if low normal it tends toward acid and if high normal it tends toward alkali.

   b. Look at pCO2 and HCO3⁻ to see if there is a disturbance corresponding to the direction suggested by the pH. If so, this is the primary disorder.

   c. Verify that there is compensation by noting that the final parameter has moved in a direction opposite to that of the primary disorder.

Summary: ABG Analysis

The key to ABG interpretation is the pH!

When the pH has drifted from a perfect 7.40, we need to find the source of the drift, which will be in either the pCO2 or HCO3⁻.

First, look at the pH and note the *direction* of the drift (acid or alkaline). Then look at the other two parameters and find the one that has moved in the *same* direction. This identifies the *primary disorder*.

Let's say that the pH has drifted in an **acid direction**. One of the other two parameters also has to have moved in an acid direction. If it's in the pCO2, we have a respiratory acidosis as the primary disorder. If it's in the HCO3⁻, a metabolic acidosis is primary.

Finally, look at the remaining (3rd) parameter. If there is movement in the 3rd parameter in the *opposite* direction, this is the *secondary* (compensatory) disorder. (Here we have an acidosis as primary, so if there were compensation, the 3rd parameter would move toward alkalosis.)

Likewise, if the pH has drifted in an **alkaline direction**, there has to be an alkalosis in one of the other two parameters. Then check the 3rd parameter to see if there is compensation (acidosis).

The "Gaps"

Two lab values that often come into play in connection with ABG interpretation are the anion and osmolar gaps.

Anion Gap

> Cation = positively charged ion in an electrolyte solution.
> Anion = negatively charged ion in an electrolyte solution.

The anion gap is the difference between the major cations (Na^+ and K^+) and the major anions (Cl^- and $HCO3^-$) in the serum. To calculate the anion gap (AG):

1. Total the cations: add the lab values for Na^+ and K^+.
2. Total the anions: add values for Cl^- and $HCO3^-$.
3. Subtract anions from cations.

Expressed as a formula, we have:

Anion Gap = Cations - Anions

– or –

AG = (Na + K) - (Cl + HCO3)

The anion gap is useful in classifying a metabolic acidosis by giving us clues about the cause and severity of the acidosis.

Some authorities prefer to calculate the anion gap *without* the K^+ cation. This is the method used in this text.

Thus, our working formula is:

AG = Sodium - (Chloride + Bicarbonate)

The approximate normal value for the anion gap is 12 mEq, and the normal range is 12 ± 2, that is, $10 - 14$.

It will be helpful to see where the normal anion gap of 12 mEq comes from. Since we said we're going to disregard the K^+ in figuring the anion gap, the only cation left is sodium. If we look up the mean value for sodium, we see that it is 140.

Thus, the total for cations is that of sodium alone, namely, 140.

The major anions are chloride and bicarbonate.

We already know that the mean value for $HCO3^-$ is 24, and looking up Cl^- we find the mean value to be 104.

Adding the 2 anions together, we get 128 for the total anion value.

$$
\begin{array}{rl}
104 & Cl^- \\
+ 24 & HCO3^- \\
\hline
128 & \text{Total Anions}
\end{array}
$$

Plugging these values into our formula $AG = Na - (Cl + HCO3)$, we have:

$$
\begin{array}{rl}
140 & \text{Total Cations } (Na^+) \\
-128 & \text{Total Anions} \\
\hline
12 &
\end{array}
$$

Thus, the major cations (Na^+) exceed the major anions by 12 mEq.

The reference range for the anion gap is 12 plus or minus 2 (12 ± 2), or about $10 - 14$.

If you wish to use K^+ in computing the anion gap, then raise the reference range values by approximately 4 mEq (the lab value for K^+).

If the anion gap is high, especially if more than 10 mEq over the normal range, the likely suspect is *acid*.

> Endogenous = originating from inside
> Exogenous = originating from outside

This acid can be *endogenous* organic acids.   Two examples are:
- ketone bodies in diabetic ketoacidosis (DKA)
- lactic acid from anaerobic metabolism (occurs in cardiac arrest).

This acid can also come from *exogenous* sources.   Some examples are:
- acetylsalicylic acid (aspirin) overdose
- poisoning from ethylene glycol (antifreeze) or methanol (wood alcohol). (These 2 toxins are metabolized into organic acids.)

If acid is added to the blood, the H^+ from the acid destroys bicarbonate. Since $HCO3^-$ is one of the 2 major anions, the sum of the Cl^- and $HCO3^-$ must get smaller, thus "widening" (increasing) the normal anion gap of 12 mEq.

$$AG = Na - (Cl + HCO3)$$

↑

IF ANIONS DECREASE AND
SODIUM STAYS THE SAME,
THE AG HAS TO INCREASE

The anion gap is helpful in differentiating a metabolic acidosis. Those who care for patients with this condition should routinely inspect the values for the electrolytes as well as the blood gases.

We can characterize a metabolic acidosis on the basis of the "width" of the anion gap.

If the anion gap is normal to slightly elevated (less than 10 mEq above the normal range), we say that this is a *normal-anion-gap* metabolic acidosis.

A metabolic acidosis whose associated anion gap is markedly above the normal range is termed a *high-anion-gap* acidosis.

A *normal*-anion-gap metabolic acidosis is most likely caused by a loss of bicarbonate such as occurs in prolonged diarrhea.

A *high*-anion-gap metabolic acidosis almost certainly comes from one of 4 sources:
- kidney failure (uremic acidosis)
- ketoacidosis
- lactic acidosis
- acidosis from drugs or toxins.

The conditions which result in a high-anion-gap acidosis are generally more acute and severe, whereas those that cause a normal-anion-gap acidosis tend to be more chronic and less severe.

Osmolar Gap

Osmolality = concentration of particles in a solution.

The serum osmolality can be directly *measured* from a blood sample. It can also be indirectly *calculated* by an equation which "counts up" the number of dissolved particles in the blood, using sodium, glucose, and BUN for the calculation.

The osmolar gap is the difference between the measured serum osmolality and the calculated osmolality. In health, the osmolar gap should be very "narrow" (low), if any.

A wide osmolar gap suggests the presence in the blood of low molecular weight solutes such as methanol, ethanol, acetone, ether, etc.

An osmolar gap of 10 mOsm or greater is considered critical.

An interesting condition occurs when there is a metabolic acidosis that is accompanied by both a high anion gap and a high osmolar gap.

The high anion gap tells us that the acidosis is probably due to a build-up of abnormal acids (rather than from a loss of $HCO3^-$) and that the metabolic acidosis is of greater acuity.

The high osmolar gap suggests some substance in the blood composed of small water-soluble molecules. Antifreeze is able to depress the freezing point of water precisely because it is composed of just such molecules.

It is very likely that the patient has ingested either ethylene glycol or methanol. Almost no other substances are capable of producing severe metabolic acidosis and a high osmolar gap.

Remember this triad:

- low pH (acidemia)

- high anion gap

- high osmolar gap.

This combination is a medical emergency requiring prompt treatment.

COMMON CAUSES,

ASSESSMENT FEATURES,

and

INTERVENTIONS

in

Acid-Base Disorders

METABOLIC ACIDOSIS

Common Causes

– Metabolic Acidosis with High Anion Gap

Most often results from accumulation of organic acids:
- Lactic acidosis from anaerobic metabolism (shock, severe hypoxia)
- Diabetic ketoacidosis (DKA)
- Poisoning (salicylates, methanol, ethylene glycol)
- Renal failure acidosis (uremia)
- Starvation ketosis.

– Metabolic Acidosis with Normal Anion Gap

Most often results from loss of HCO_3^-:
- Diarrhea
- Uretero-enterostomies
- Pancreatic fistula
- Renal tubular acidosis (RTA).

Assessment

- Check vital signs for shock ($\downarrow$ P & BP).
- Assess for CNS depression (lethargy, delirium, coma).
- Observe for signs of dehydration (dry mucous membranes, "tenting" of skin, sunken eyeballs, $Na^+ > 150$).
- Hyperventilation if pt. is compensating the acidosis.
- Monitor electrolytes, esp. Na^+ and K^+ (latter tends to be high in acidosis).
- Calculate the anion gap (the wider the gap, the sicker the pt.).
- Arrhythmias may develop (irregular pulse).
- Urine may become quite acid (urine pH < 4.5).
- Look for wide osmolar gap in suspected poisoning and/or coma.

METABOLIC ACIDOSIS

Interventions

- If pt. is in cardiac arrest, begin CPR.
- If not in arrest, keep siderails up.
- Treat shock (keep warm, supine).
- Insure patent airway and adequate oxygenation.
- Monitor IV fluid therapy to support circulation, maintain hydration and electrolyte balance.
- If pH < 7.2 or pt. is in long arrest (> 10 min), candidate for sodium bicarbonate IV push.
- If pt. is in DKA, major modalities include insulin, K^+, large volume of IV fluid.

RESPIRATORY ACIDOSIS

Common Causes

Always involves impaired alveolar ventilation:
- COPD and other respiratory diseases
- Pulmonary edema
- Inadequate mechanical ventilation
- Excess CNS depressants
- Head trauma
- Neuromuscular disease.

Assessment

- There may be marked respiratory depression and/or dyspnea (for example, wheezing).
- Assess for CNS depression (disorientation, stupor).
- Arrhythmias are possible (irregular pulse).
- Urine pH may fall < 4.5.

Interventions

- Improve ventilation.
- Client safety (siderails up).

MIXED ACIDOSES

Common Causes

Most frequent is cardiopulmonary arrest. Others are severe pulmonary edema and drug ingestion with CNS depression.

Assessment

As previously noted under Metabolic Acidosis and Respiratory Acidosis.

Interventions

- If arrest: CPR, life support drugs.
- If pulmonary edema: bedrest with head of bed elevated, oxygen, meds (diuretics, digitalis, morphine, afterload reducers), Na^+ restriction.
- If drug O.D.: supportive care.
- Mechanical ventilation may be needed or rate may need to be increased.

METABOLIC ALKALOSIS

Common Causes

- Loss of Cl^-:
  - » Vomiting
  - » Diuretics
  - » NG suction.
- Intake of alkali (e.g., sodium bicarbonate).
- Steroid excess (Cushing's syndrome).
- Severe K^+ deficiency (hypokalemia).
- Occasional causes to watch for:
  - » Infusion of large volume of blood (citrate is metabolized to $HCO3^-$).
  - » Sodium-salt penicillins (Na^+ is reabsorbed in tubule, but H^+ & K^+ are excreted, resulting in hypokalemic alkalosis).

Assessment

- Assess for CNS signs such as disorientation, lethargy, convulsions.
- Muscle irritability may produce twitching and tetany.
- There may be arrhythmias, esp. if K^+ is very low (irregular pulse).
- Urine may become alkaline (urinary pH > 7.0).

Interventions

- Increase extracellular fluid volume.
- Administer K^+ and Cl^- replacement.
- Protect client (siderails up).

RESPIRATORY ALKALOSIS

Common Causes

Any condition that results in hyperventilation:
- Anxiety
- Pain
- Hypoxemia from lung disease
- Hypermetabolic states (fever, D.T.'s, hyperthyroid)
- Excessive mechanical ventilation
- Gram-negative sepsis
- Kussmaul's respirations in DKA
- Overstimulation of respiratory center in brain by ASA poisoning.

Mild respiratory alkalosis which usually does not require treatment is found in pregnancy, cirrhosis, and CNS lesions.

Assessment

- CNS signs that are characteristic of respiratory alkalosis are lightheadedness, dizziness, disorientation, convulsions.
- Typical peripheral nervous system signs are paresthesias and hyperactive reflexes.
- Observe for muscle irritability (twitching, tetany).

Interventions

- Treatment is aimed at the underlying disorder.
- If pt. is on a ventilator, rate may need to be decreased.
- Use appropriate measures to reduce anxiety, pain, fever.
- Insure open airway, adequate oxygenation.
- Keep siderails up.

- Rebreathing CO_2 in a paper bag has been used to treat respiratory alkalosis, but this may induce *hypoxemia*.

MIXED ALKALOSES

Common Causes

- Liver failure + diuretics or NG suction.
- Ventilator patients given NG suction.

Assessment

As previously noted under Metabolic Alkalosis and Respiratory Alkalosis.

Interventions

- For metabolic alkalosis, administer volume, Cl^-, and K^+ replacement.
- For respiratory alkalosis, decrease ventilator rate.
- Patient security (siderails up).

Quiz

1. **The normal value for the pH is:**
   a. 7.00
   b. 7.44
   c. 7.40
   d. 4.70

2. **The normal value for the pCO2 is:**
   a. 40
   b. 14
   c. 7.4
   d. 4

3. **The normal range for the bicarbonate (HCO3⁻) is:**
   a. 35 - 45
   b. 7.35 - 7.45
   c. 8 - 12
   d. 22 - 26

4. **The metabolic parameter of the ABG set is the:**
   a. pH
   b. pCO2
   c. HCO3⁻
   d. BUN

5. **The respiratory parameter of the ABG set is the:**
   a. FEV_1
   b. pO2
   c. HCO3⁻
   d. pCO2

6. **Acidemia would be indicated by:**
   a. pH less than 7.35
   b. pCO2 greater than 45
   c. HCO3⁻ more than 26
   d. serum osmolality less than 280

7. **Alkalemia would be indicated by**:
   a. HCO3⁻ less than 22
   b. pH less than 7.35
   c. pCO2 less than 35
   d. pH more than 7.45

8. **The acid-base disorder associated with excessive ventilatory movements is**:
   a. lactic acidosis
   b. respiratory alkalosis
   c. high anion gap metabolic acidosis
   d. hyperventilation acidosis

9. **An abnormally large accumulation of ketones in the blood would give rise to a(n)**:
   a. metabolic acidosis
   b. pH greater than 7.45
   c. metabolic depression
   d. respiratory acidosis

10. **Respiratory failure is often associated with a(n)**:
   a. low anion gap
   b. elevated pO2
   c. depressed pCO2
   d. pCO2 above 45

11. **In a long-standing metabolic acidosis, the respiratory rate may increase. This is an example of**:
   a. mixed acidoses
   b. a low osmolar gap
   c. compensation
   d. a primary respiratory acidosis

12. **The difference between the major cations and the major anions in the serum is called the**:
    a. decompensation factor
    b. osmolar gap
    c. degradation product
    d. anion gap

13. **The presence of low molecular weight solutes such as ethylene glycol in the blood is most likely to cause a(n)**:
    a. osmolar gap
    b. respiratory acidosis
    c. low anion gap
    d. metabolic alkalosis

14. **The approximate normal value for the anion gap is**:
    a. 4.7
    b. 4.0
    c. 12
    d. 24

15. **A base excess of -12 would indicate**:
    a. mixed alkaloses
    b. respiratory alkalosis
    c. metabolic acidosis
    d. a wide osmolar gap

16. **An S_pO2 of 85 would indicate**:
    a. hypoxemia
    b. shift to the left
    c. respiratory alkalosis
    d. adequate oxygenation

17. **The osmolar gap suggests the presence in the blood of**:
    a. exogenous bicarbonate
    b. major cations
    c. low molecular weight substances
    d. desaturated hemoglobin

18. **Mr. X is complaining of dizziness and a tingling sensation in the extremities. He is confused and hyperventilating. His signs and symptoms are most consistent with**:
    a. COPD
    b. dehydration
    c. mixed acidoses
    d. respiratory alkalosis

19. **A capnogram gives us information about the client's**:
    a. acid-base status
    b. status of ventilation
    c. electrolytes
    d. status of oxygenation

For items 20–26, select the most probable interpretation for the given ABG set.

20. **pH 7.51, pCO2 40, HCO3⁻ 31** :
    a   normal — no acid-base disorder evident
    b. uncompensated metabolic alkalosis
    c. partially compensated respiratory acidosis
    d. mixed acidoses

21. **pH 7.36, pCO2 29, HCO3⁻ 16** :
    a. partially compensated respiratory alkalosis
    b. uncompensated metabolic acidosis
    c. normal — no acid-base disorder evident
    d. fully compensated metabolic acidosis

22. **pH 7.40, pCO2 40, HCO3⁻ 24** :
    a. secondary metabolic alkalosis
    b. mixed alkaloses
    c. mixed acidoses
    d. normal — no acid-base disorder evident

23. **pH 7.12, pCO2 60, HCO3⁻ 19** :
    a. mixed acidoses
    b. partially compensated respiratory acidosis
    c. fully compensated metabolic alkalosis
    d. normal — no acid-base disorder evident

24. **pH 7.48, pCO2 32, HCO3⁻ 23** :
    a. uncompensated respiratory alkalosis
    b. normal — no acid-base disorder evident
    c. uncompensated respiratory acidosis
    d. uncompensated metabolic alkalosis

25. **pH 7.62, pCO2 30, HCO3⁻ 30** :
    a. fully compensated respiratory acidosis
    b. mixed alkaloses
    c. normal — no acid-base disorder evident
    d. partially compensated respiratory alkalosis

26. **pH 7.30, pCO2 59, HCO3⁻ 28** :
    a. normal — no acid-base disorder evident
    b. partially compensated metabolic acidosis
    c. mixed acidoses
    d. partially compensated respiratory acidosis

— End of Quiz —

Answers to Quiz

1.   c
2.   a
3.   d
4.   c
5.   d
6.   a
7.   d
8.   b
9.   a
10. d
11. c
12. d
13. a
14. c
15. c
16. a
17. c
18. d
19. b
20. b
21. d
22. d
23. a
24. a
25. b
26. d

Companion Software

If you enjoyed this book, a complete computerized learning program for blood gases is available. This program, *A-B-Gee!*™, contains the following modules:

- **Tutorial**. Content is similar to material in this book, except that learner is coached in bite-size steps and program periodically checks understanding to keep you on track. You control pace of learning, and a variety of help is always available. Select entire tutorial or any topic from menu.

- **ABG Interpreter**. Validating ABG sets from actual clinical reinforces the learning process. Practice by obtaining values from lab reports. Make a tentative interpretation, then enter values and compare your interpretation to program's analysis. Click "Explanation" button and program walks you through each step of analysis. Option to view common causes, assessment parameters, and interventions for identified acid-base disorder.

- **Guess A Gas**. A game to hone your interpretation skills. A few minutes every six months will keep them sharp.

- **Pump A Gas**. Helps you visualize effect that varying pCO2 and/or bicarbonate has on pH. "Pump" these parameters up or down and see how the interpretation changes.

- **Interactive Quiz**. Gives immediate explanation on incorrect choices that guides you to correct answer.

Available on CD for Windows XP or later. Send inquiries to dthompson16@nc.rr.com. To purchase, send check or money order for $19.95 USD (outside US add $5) to:

Donald Thompson
P. O. Box 61117
Raleigh, NC 27661-1117 USA

Made in the USA
Coppell, TX
15 January 2024

27726514R00032